Rice

Anne Mountfield

Blackwell Education

Text copyright © Anne Mountfield 1989
Illustrations copyright © Basil Blackwell Ltd 1989

First published 1989
Basil Blackwell Ltd
108 Cowley Road
Oxford OX4 1JF
UK

British Library Cataloguing in Publication Data
Mountfield, Anne
  How food begins.
  1. Food – For children
  I. Title   II. Series
  641.3

  ISBN 0–631–90338–0
  ISBN 0–631–16636–X: v.6: cased
  ISBN 0–631–90337–2: v.6: pbk

Illustrated by Corinne Burrows
Typeset in 20 on 30 pt Century by
Columns of Reading
Printed in Great Britain by MacLehose & Partners, Portsmouth

Contents

Do you like rice with meat?

... or rice with vegetables?

... or rice in a milk pudding?

Guess what?

Rice begins . . .

. . . as a plant under water!

This farmer grows rice for his family to
   eat.
He is getting ready to plant the rice.
A water buffalo pulls the plough.

The farmer has made mud walls round
   his rice field.
The walls hold in rain and river water.

The farmer plants the rice seeds in
    one corner of the field.
The young plants take a month to grow.

Planting them out is muddy work.
All the family help.

Some rice farmers live in hilly places.

They cut rice fields into the side of a
   hill.

The hill looks like a giant staircase.

This rice farm is in America.
The farmer grows rice for other
    people to eat.
This rice will go to the shops.
A plane flies over the rice field.

It scatters the seed on the field.

The flag shows which way the wind is
blowing.

It tells the farmer where the
seeds will fall.

Rice plants look like tall grass.

Each plant makes seeds called grain.

ripe rice

When the plants turn yellow the
rice is ready to pick.

These farmers cut the rice with knives.
Their big hats are made of rice straw.
The hats keep off the sun.

The rice is left to dry.

Then the farmers hit the stalks on a
wooden board.
This makes the grain fall off.

Some farmers use combine harvesters to
cut the rice and pull off the grain.

First they must let out the water and
    leave the fields to dry.
If they left the fields muddy, the
    combine harvesters would get stuck in
    the mud.

The family put their rice in a shed.

They build the shed high up, to keep the
rice dry.

Animals cannot get in to eat it.

husk

bran

This rice is waiting to go to the mill.
Machines will take the hard husk off.
They take the brown bran off to
    make white rice.

Long grain rice

Brown Rice

Rice

rice
pudding

long grain rice

brown rice

short grain rice

There are many kinds of rice.

pilau rice

boiled rice

rice salad

fried rice

rice pudding

We can cook rice in many ways.

Which do you like?

Index